W9-DGT-782

SENSIBLE CLOSET FIXTURE.

THE HISTORY OF EVERYDAY LIFE

Elaine Landau

TFCB

TWENTY-FIRST CENTURY BOOKS

Minneapolis

Twenty-First Century Books
A division of Lerner Publishing Group
241 First Avenue North
Minneapolis, MN 55401 U.S.A.

Website address: www.lernerbooks.com

Library of Congress Cataloging-in-Publication Data

Landau, Elaine.
   The history of everyday life / by Elaine Landau.
      p.   cm. — (Major inventions through history)
   Includes bibliographical references and index.
   ISBN-13: 978-0-8225-3808-0 (lib. bdg. : alk. paper)
   ISBN-10: 0-8225-3808-3 (lib. bdg. : alk. paper)
      1. Inventions—History. I. Title. II. Series.
   T15.L18 2006
   609—dc22   2005009415

Manufactured in the United States of America
1 2 3 4 5 6 – DP – 11 10 09 08 07 06

CONTENTS

Introduction

You're lucky! You live in an age of countless machines, gadgets, and laborsaving innovations that make your life pretty easy.

What? You don't think your life is so easy? After all, you have lots of chores. You have to clean your bedroom, help fix dinner, fold the laundry. . . .

But imagine what life was like for kids in the past. Not that long ago, central heating, indoor plumbing, automatic appliances, and store-bought goods were luxuries many people didn't have. Many families, including the kids, spent their days chopping wood, hauling water, washing clothes by hand, sewing, and cooking food from scratch.

Yes, you are lucky. There are many inventions you probably never think about. But your life would be very different without them. If you're curious about how things were done in the past and how these inventions changed all that, just turn the page.

CHAPTER 1

Fireplaces and Central Heating

Imagine taking a trip back in time. Stop at the Middle Ages, around the year A.D. 1000. You live in a European castle. Not bad, you think. Who could complain about royal living? But wait a minute. It's the middle of winter and the temperature has dropped below freezing. You're cold, and the fur blanket you've wrapped around yourself is not really helping.

The Middle Ages
A.D. 500–1500

If this were modern times, you could simply turn on the heat. But in the year 1000, central heating is unimaginable. In fact, not even a simple fireplace has been invented yet. The only source of heat is an open fire pit surrounded by stones on the first floor of the castle. It looks like a campfire that's been moved indoors. The smoke is supposed to escape through a hole in the roof. Yet that hardly does the job—it's pretty smoky at your castle. And unless you stay right in front of the fire, it's also quite cold and damp.

> ### VESTA
> The ancient Romans had a goddess of the hearth, or fireplace, named Vesta. This goddess was highly honored in all households.

Up in Smoke

By the 1250s, castle life had improved somewhat. Most ordinary homes still had fire pits. But indoor fire pits had been replaced in castles with permanent fireplaces. Most castles had a huge fireplace in the kitchen. Servants cooked food over the fireplace in pots hanging from chains above the fire.

Still, there was no efficient way to route the smoke out of the building, and castles remained pretty smoky. Then, in the 1500s, chimneys came into use. The warm smoke rose up into a hood directly above the fire and was channeled out of the building

Fireplaces are common in European castles.

1250s

An Englishman and his dogs bask beside a roaring fire. The warmest place in most homes was next to the fireplace.

through the chimney. Yet early chimneys were not very efficient compared to modern chimneys. Much of the heat went up the chimneys with the smoke. A nearly equal amount of smoke spilled into the room.

By this time, average homes had fireplaces too. They were used for warmth, cooking, and light for reading and sewing. They were kept burning all day. Someone had to constantly tend the fire to make sure it didn't go out. Back then people didn't have matches or lighters. It was easier to keep the same fire going than to start a new one. At night they often let the fire burn down. Then they stoked the embers to relight the fire in the morning.

With chimneys, fireplaces become common in average homes.

1500s

In Early America

Colonists and pioneers in early America built fireplaces too. These were much larger than modern fireplaces. Many were between six and ten feet (two to three meters) wide. Whole families gathered around the fireplace on chilly winter nights. Early American families ate, slept, and spent most of their free time together in front of the fire. That's because other parts of the home remained very cold. And social gatherings were uncommon. People didn't want to get together outside their homes, where there wasn't enough room next to a fireplace for everyone to keep warm.

MORE THAN TOASTING MARSHMALLOWS

Before there were stoves, colonial cooks used fireplaces. A cook had to be able to control the fire and know which woods to use. Hardwoods such as ash, oak, hickory, and dogwood were best. These gave off an even and intense heat. A blazing fire was great to look at but could badly burn a stew. A small fire was needed for roasting and toasting. On rainy days, the cook had to bring the wood supply inside. The logs were stacked around the sides of the fireplace, so the fire's heat dried them out.

Inventors and scientists such as Benjamin Franklin and Benjamin Thompson created increasingly safe and efficient fireplaces and ovens. Builders gained confidence in their skills and in these scientific developments. They began to build homes with fireplaces in more than one room. Finally, family members could

Ben Franklin designs his first iron fireplace.
1740s

The Franklin stove was a cast-iron box that was fitted into the fireplace. It extended into the room so that three sides gave off heat instead of only one.

sleep in private rooms. Still, people couldn't get too comfortable. Fireplaces still didn't provide much heat unless you were right next to them.

In addition, fireplaces were a lot of work. Wood needed to be chopped and hauled, the fire needed to be tended, and ashes needed to be cleaned out. It was messy work. Even the best colonial housekeepers fought a daily battle with dust and ashes.

The Basement Furnace

Things warmed up in the mid-1800s, when the first homes were built with centrally located chimneys. Previously, chimneys had always been built along an outside wall. But a lot of heat was lost through the back of the chimney. With a centrally located chimney, one fireplace could heat a home much more efficiently.

Chimney design improves when they are centrally located in houses.

1800s

By the end of the Civil War (1861—1865), many wealthy families in the United States and Europe had a basement furnace. It was a large coal- or wood-burning oven that heated fresh air and pushed it up through ducts, or tubes, into upstairs rooms. This was central heating, as the heat came from one central source. By 1950 most homes in the United States had central heating. Most modern central heating systems run on electricity or burn natural gas instead of coal or wood.

Patrons at a restaurant in Paris in the late 1800s could enjoy the comfort of dining in centrally heated spaces.

Wealthy families have
basement furnaces.

1860s

Most U.S. homes have
central heating.

1950s

Putting fireplaces in more than one room of a house had given people some privacy. But central heating did even more. Houses could be built larger and heated better. All activity no longer needed to be centered in one location. Children had their own bedrooms. Architects even began designing houses with separate recreational areas or game rooms.

Central heating changed people's lives outside the home as well. Families did not need all their energy just to survive the brutally cold winters. People no longer waited till the snow thawed each spring to get together. Social events could be held during the cold winter months. Centrally heated theaters, banquet rooms, and concert halls became part of the American way of life. Social and cultural activities continued to expand. You can comfortably go to centrally heated movie houses, restaurants, museums, and indoor sports arenas.

Of course, some families still like the idea of gathering around a fireplace. Many people yearn for an earlier time, when families spent all their evenings together. For these people, the fireplace is a symbol of that era. Yet with central heating, families can also

ARCHITECTS KNOW BEST!

The famous American architect Frank Lloyd Wright (1867–1959) once said, "Any room of importance has a fireplace." Wright designed his houses to be peaceful retreats for families. He often made a large fireplace the central focus of a room, a place to draw people together.

This family enjoys some time together in front of their fireplace. Thanks to the invention of central heating, the fireplace is a luxury, not a necessity.

go out and do fun things together. Central heating gives us a wide range of choices. People have fuller lives today because of it. Fortunately, the days of huddling around indoor fire pits are long gone.

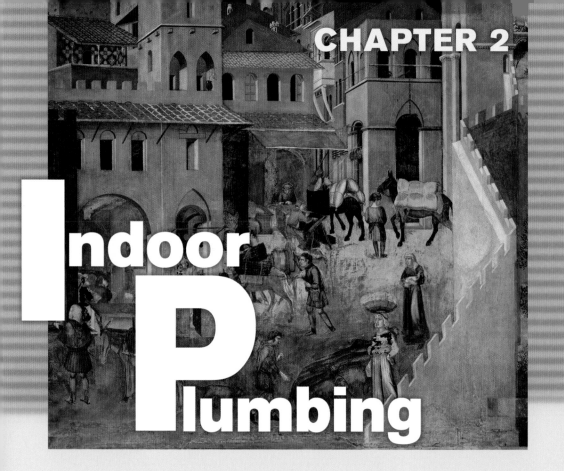

CHAPTER 2

Indoor Plumbing

Good thing you didn't live in Rome, Italy, in the 1300s. Just walking down the street could be dangerous. Back then, houses didn't have indoor toilets. Instead, people used ceramic containers called chamber pots. Each bedroom had a chamber pot, and servants had to empty the pots regularly. Most of the time, they just tossed the contents out the front door or from an upper-story window.

Sometimes people walking down the street had to be quick and

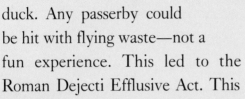

A chamber pot
from the 1600s

duck. Any passerby could
be hit with flying waste—not a
fun experience. This led to the
Roman Dejecti Efflusive Act. This
law gave people injured by flying waste the right to collect money
from whoever flung it.

Keeping Clean in the Ancient World

Long before chamber pots and open windows, people had looked
for ways to get rid of natural waste. The first humans used naturally
flowing water bodies to carry away their waste. They got drinking
and bathing water from these same rivers and streams. A people's
system for distributing and using water—both coming and going—
is its plumbing system.

The first-known indoor bathrooms appeared about 5,000 years
ago, on the Orkney Islands, north of Scotland. Stone huts discovered
there contain hollowed-out spots along the walls, with drains dug into

Orkney islanders build
indoor toilets.
3000 B.C.

The ancient stone huts on the Orkney Islands overlook the North Atlantic Ocean. Archaeologists found the earliest-known indoor toilets here.

an underground sewer system. Rich people in ancient Babylon (modern-day Iraq) had bathing rooms with waterproof floors that sloped toward a drain. The drain led to a sewer system. Slaves poured clean water over a bather, and then the water flowed into the drain.

Many other ancient cultures developed underground sewer systems, bathtubs, showers, and toilets. The ancient Egyptians even had hot and cold running water and copper pipes. The earliest

Egyptians use
sitting-type toilets.

2100 B.C.

Most people continued to get rid of household waste out the nearest door or window. If the waste on your street and in your water didn't give you a terrible disease, you still had to live with the smell. In London, England, waste was dumped into the Thames River. The Thames became a receptacle for human waste, animal carcasses, food refuse, and even dead bodies. In the summer months, the river smelled so bad that many of London's wealthier citizens gladly fled the city for their country estates. The not so wealthy had to tough it out.

THE GREAT STINK

The Houses of Parliament, the seat of Great Britain's government, are located on the east bank of the Thames River in London. Like all Londoners, members of Parliament had to deal with the odor from the Thames. Often heavy curtains coated in lime, a powdered mineral, were hung in Parliament's windows to keep the smell out. But in 1858, the stench from the river grew so overwhelming that the members of Parliament abandoned the buildings for a time. Soon after the Great Stink of 1858, large sewers were built on both sides of the river.

The Outhouse

Conditions during pioneer days in the United States were not much better. Modern bathrooms were still unheard of. Pioneer families had an outhouse—a small wooden hut built over a hole in the ground. Inside, people sat on a seat with an opening. The waste dropped through the opening into the pit below.

BATH TIME

Taking a bath in pioneer days was quite a chore. Water had to be hauled in from a well, heated over the fireplace, and poured into a tub. That was a lot of work, so Americans on the frontier did not bathe often—usually once a week. When they did, the whole family often took turns using the same bathwater. They usually went in order of age. If you were the youngest child, you were the last to bathe. By then the water was probably pretty dirty and cold.

Luckily, indoor plumbing and water heaters changed all that. With a water heater, water coming out of a faucet was already hot. The hot water could be used for washing clothes, doing dishes, and cooking too.

Outhouses weren't very comfortable. On freezing winter nights, the seats felt cold, even icy. And who would want to walk outside to the outhouse in your pajamas in the middle of winter? The summer was bad too. Outhouses were hot, stinky, and full of flies.

Modern Sanitation

Sewage continued to flow in the streets and rivers of European cities, and people continued to get sick and die from diseases such as cholera. Then, in the 1850s, an English doctor named John Snow proved that drinking water polluted with waste was causing cholera. Huge sanitary reforms were made after that, preventing future outbreaks of the deadly disease. Sewers were built all over London, and laws were passed requiring houses to have flushing toilets.

Around the same time, modern sewage systems became common in the United States too. In 1857 an engineer named Julius Adams de-

A severe cholera outbreak sweeps through London.

1854

signed a sewer system for Brooklyn, New York. Other large cities came up with systems to carry out waste as well. Indoor plumbing paved the way for modern bathrooms. With running water, bathrooms complete with toilets, sinks, bathtubs, and showers were possible.

But indoor plumbing and proper sewerage systems led to more than pretty rooms and better smelling people. It also gave a big boost to public health. Diseases such as cholera and dysentery no longer wiped out large numbers of people. Instead, the diseases themselves were nearly wiped out. Medical advances have been important for fighting disease. Yet society also owes a debt to the plumbers and engineers who designed indoor plumbing. They improved sanitation for everyone.

THE TRUTH ABOUT THOMAS CRAPPER

Thomas Crapper was a plumber who lived in England in the 1860s. Some people wrongly believe that Crapper invented the toilet. Actually, Crapper's plumbing company merely manufactured toilets. During World War I (1914–1918), U.S. soldiers passing through England saw Crapper's name on the toilet tanks made by his company. The soldiers began using the slang word "crapper" to mean toilet. Over time, people came to believe that Crapper invented the toilet.

Modern plumbing has solved many health and convenience problems. Pipes bring fresh water into your home for drinking, bathing, and cooking. Used water and waste go down a drain and out of your house through drainpipes. You hardly have to think about it.

Julius Adams designs a
sewer system for
Brooklyn, New York.
1857

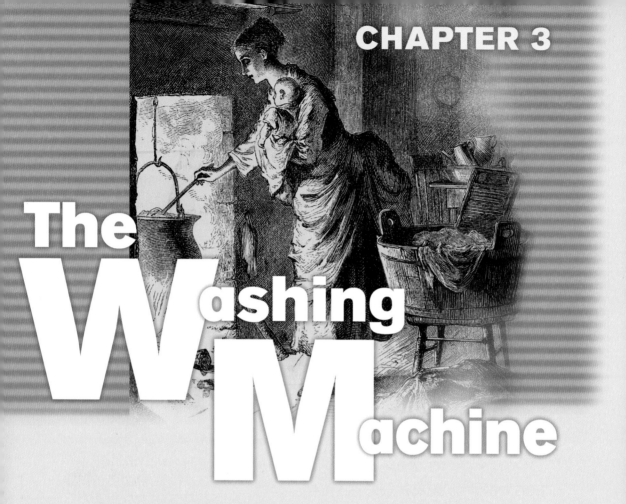

CHAPTER 3

The Washing Machine

Picture this: You've got a soccer game this afternoon. Up until a few minutes ago, you hadn't thought about your uniform. Suddenly, you remember that you left it crumpled in the back of your closet. It looks filthy and smells worse. You call out, "Mom! Dad!" But there's no answer. Nobody will be home until it's time to drive you to the game. You're in a fix—but fortunately the answer is just steps away, in the laundry room. With a washing machine and dryer, your soccer uniform will be clean with time to spare.

This would not have been the case if you lived in ancient times. You wouldn't have been on a soccer team, true, but you'd still need clean clothes. And doing the wash wasn't easy back then. Ancient people carried their dirty clothes down to the nearest stream. There, they pounded them on rocks to get them clean. It was a difficult and tiring job, one that was usually left for women to do. For most of history, laundry has been women's work.

Laundry Day

Laundry continued to be women's work in early America. According to diaries kept by pioneer women, washing clothes was the chore they dreaded most. In many pioneer families, Monday was laundry day. So on Sunday evenings, the women began hauling in buckets of well water. If there wasn't a well, they carried in buckets of rainwater that had collected in an outdoor trough. When there hadn't been much rain, the women fetched the water from streams. In some cases, the nearest stream was a long walk away.

After collecting the water, they heated it over the fireplace. The women soaked the dirty clothes in warm water overnight. Early the next morning, they'd begin scrubbing the items.

Scrubbing was done on a washboard. Some people regard washboards as the earliest washing machines. Invented in 1797, the

The washboard
is invented.
1797

washboard was a wooden board with grooves and ridges on its surface. To get the dirt out, women rubbed the clothes against the board. For ground-in dirt, they used a soap made of lye. This soap was strong enough to take paint off a building. It did the job of cleaning clothes, but it stung and burned skin. Often, the women's hands were left red and irritated.

After scrubbing the clothes, women placed them in a big kettle of boiling water. They used a large pole called a wash stick to move the clothes around. They used the same wash stick to lift the clothes out and put them in rinse water.

Finally, women carried the heavy, wet clothes out to dry. Each water-laden piece was wrung out by hand. Then the clothes were hung on a clothesline. Some women liked to place the items over pleasantly scented hedges to dry.

Doing laundry this way was exhausting and took a great deal of time. It was a woman's main focus from Sunday night to Monday

CHINESE LAUNDRIES

During the 1850s, many American men headed west hoping to strike it rich in the California gold rush. Many men came from other countries too, including China. In America the Chinese faced a great deal of prejudice. They were not allowed to prospect for gold except on worthless claims that had been abandoned.

Needing to make money, the Chinese found other work. The prospectors and miners needed to have their clothes washed, at least occasionally. The Chinese saw this need and began opening laundries. In the mid-1800s, the Chinese owned and operated most of the laundries in California.

Chinese laundries
open during the
California gold rush.

1850s

night. Of course, doing laundry did not free her from other chores, such as making meals.

The First Washing Machines

There had to be a better way to get the job done, and a number of inventors were anxious to find it. Among the first early washing machine inventors was an Indiana corn planter and manufacturer named William Blackstone. He created a washing machine as a birthday gift for his wife in 1874.

Blackstone's washer was a hand-operated machine that looked like a wooden tub with wooden pegs inside it. To wash clothes, you filled the tub with hot, soapy water and dirty clothes. Then you turned a handle. The dirty clothes caught on the pegs and moved through the water until they were clean. But sometimes the pegs tore the clothes.

William Blackstone builds one of the first washing machines.

1874

This woman uses an early motorized washing machine in the 1920s. These machines included wringers, which squeezed excess water out of the clothes so women didn't have to do it by hand.

By about 1900, the first electric washing machines were developed. A motor moved the tub. As the tub moved, the soapy water was pushed through the clothes, lifting out the dirt. But these early machines needed some improvements. At first, the motors were not properly shielded from the wash water. Water dripped on the motors, causing short circuits. Women using these machines sometimes got jolting electric shocks.

In 1922 the Maytag Corporation began producing agitator washing machines. These new machines were a big improvement

Electric washers
are developed.
1900

The Maytag Corporation
produces agitator washers.
1922

over previous designs. They had a cone-shaped part called an agitator in the center of the tub. The agitator's action moved the clothes back and forth. It also moved water through the clothes. The public welcomed these new washing machines.

Automation

During the 1960s, automatic washing machines became the rage. They promised to make laundry day amazingly simple. The user only had to set a few knobs. Everything else, such as controlling the water's temperature, was automatic. The user could even adjust the strength of the wash cycles. The delicate cycle could be set for items such as blouses and underwear. The heavy-duty cycle could handle denim work clothes or blankets.

Over the years, washing machines came into widespread use in the United States. Even people who don't own their own washers

THIS IS THE WAY WE DRY OUR CLOTHES

Clothes dryers were nearly as important as washing machines. Outdoor clotheslines were useless on rainy days. At those times, an indoor clothesline had to be put up. This was difficult unless you had plenty of extra space in your house. The first electric clothes dryers were developed around 1915. Since then, clothes dryers, like washing machines, have improved. But the idea behind them remains the same. Warm air is circulated in these machines to dry damp clothes. Few people use washing machines without using dryers too.

Automatic washers hit the market.
1960s

These kids help out around the house by doing laundry.

can use coin-operated machines in Laundromats or laundry rooms in their apartment buildings.

The washing machine gave women a huge part of their lives back. What once took a whole day every week can be done in a couple of hours. What once was a massive chore for women is a small task for anyone with a few minutes—including kids.

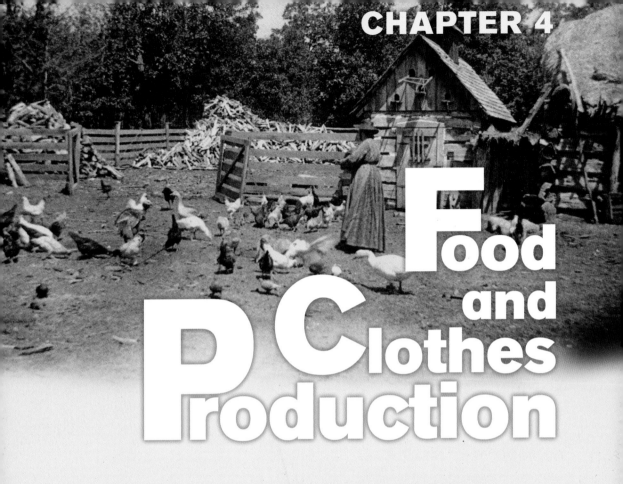

CHAPTER 4

Food and Clothes Production

American pioneers didn't buy ready-made clothes or fresh food from stores. There was little variety in what they wore and ate. Most families lived in the country. Many lived on the frontier—the western edge of U.S. territories. Their cabins were often a day's drive from the nearest town and its shops.

These families had to grow or make most of what they needed. That meant long days of backbreaking work. Even people who lived in cities raised chickens and cows and grew vegetables in a garden. And they cared little about fashion. They did not wear designer labels. They didn't have different shoes for different outfits or go to the malls to check out the stores. In early America, there were no malls. Few stores sold clothes or food.

Putting Food on the Table

Before refrigerators and fast transportation, such as trains, Americans ate what they could find around them. There was no way to bring food from other parts of the country before it spoiled.

So people who lived by the water ate a lot of fish. People who lived in the South ate a lot of sweet potatoes. Most farmers had plenty of corn, which could be eaten fresh or ground into meal and baked into bread.

Men hunted wild animals, such as buffalo, deer, wild turkeys, and quail for food. Pioneer families raised farm animals, such as pigs and chickens, as well. Pork could be salted and preserved through an entire winter, even without a refrigerator. People grew hardy vegetables, such as potatoes, turnips, pump-

BEWARE OF BEARS AT BERRY PATCHES!

Berry picking on the frontier could be risky. That's because bears also like berries. Berry pickers always went out in pairs. One did the picking while the other remained on the lookout for uninvited guests.

John Kay invents the flying shuttle for weaving looms.

1733

James Hargreaves invents the spinning jenny, which spins multiple spindles of thread.

1764

kins, and beans. Leafy vegetables, such as spinach, spoiled too easily. People often explored the nearby wooded areas for wild berries, wild crab apples, and nuts.

Keeping Food Fresh

People could get a little variety by trading food with neighbors or buying food at a town market. But people didn't buy much food because it didn't last long. Some foods could be canned at home. Fruits and vegetables grown in family gardens over the summer were cooked and sealed in clean glass jars. Families could eat the canned food all winter.

Keeping meat and dairy products fresh was a bigger problem. But in 1803, the first icebox patent (a claim to legal rights for an invention) was issued. An icebox was like a modern

A young man places ice in an icebox.

Pioneers move westward after the American Revolution.
1790s

The first icebox patent is issued.
1803

refrigerator, but it didn't use electricity to keep food cool. Instead, a big block of ice was kept inside. Iceboxes were expensive, though, and so was ice. Then, in 1827, an ice cutter was invented to efficiently harvest ice from ponds and lakes. That brought the price of ice down and made iceboxes much more common.

Things began to change in the 1870s. By then refrigerated railroad lines brought fresh food to consumers all over the United States. Early on, this only benefited rich people. It was expensive to haul fresh food a long distance, and few could afford to buy it.

Commercial Food Production

About the same time iceboxes were becoming common, commercial canning became popular. H. J. Heinz started to bottle and sell horseradish, pickles, and sauerkraut in Pennsylvania. Within the next ten years, his company added cooked macaroni to its product line.

The Franco-American Food Company of New Jersey came on the scene with entire meals in a can. All you needed was a few minutes and a can opener to enjoy a tasty spaghetti dinner. Customers just opened the can, poured its contents into a pan, and heated it on the stove. Preparation was so simple that even young people could make their own lunch or dinner.

As time passed, canned goods increasingly became a part of our lives. In 1860 just 5 million cans of food were produced in the United States. But in 1870, Americans bought 30 million cans of

The ice cutter improves ice harvesting from ponds and lakes.

1827

Elias Howe patents the first sewing machine.

1846

An advertisement from 1896 for Franco-American canned soups—ready for use!

FRANCO-AMERICAN SOUPS
READY FOR USE
CLEAN, HONEST, APPETIZING.
Sold by Grocers Everywhere.
FRANCO-AMERICAN FOOD CO.

these products. By 1880 that number had soared to 120 million cans. By that time, a can opener was considered a necessary kitchen tool.

Canning food was just the start. In the 1900s, all sorts of processed and prepared foods found their way into America's pantries and cupboards. There were birthday cake mixes, packages of cookies, crackers, pretzels, and boxes of candy.

The General Electric Company unveiled two of the first home refrigerators in 1911. People no longer had to buy ice, which was often cut from polluted water, for their iceboxes. Refrigerators kept fresh food cool with electricity. Before long, Americans could choose from more than two hundred refrigerator models.

Electric sewing and cutting machines speed up clothes production.

1900s

General Electric begins selling home refrigerators.

1911

REFRIGERATORS AND SUPERMARKETS: PERFECT TOGETHER

Home refrigeration changed how we lived in many ways. With refrigerators and freezers at home, people could buy more food and keep it longer. That meant they didn't have to shop as often. But the small grocers and markets common in most neighborhoods didn't have much variety. To really take advantage of refrigeration, people needed larger stores with lots of different foods— fresh meat, fish, dairy and bakery items, and frozen foods. It would have to be a really super market.

By the 1920s, supermarkets dotted America's landscape. They carried just about every kind of food people wanted and, because they sold in large volume, they could offer low prices.

By the 1940s, refrigerators with separate freezer compartments were sold. People could make their own ice. They could also eat frozen meats, fish, fruits, and vegetables from all over the country, even all over the world.

All these foods were no longer limited to the rich. Transportation, refrigeration, and canning had created a need for mass production. Mass production is the manufacturing of a product in large numbers, or quantities. Companies such as Franco-American could make huge amounts of their product and ship it all over the country. Because they made so much, it cost less to make. That meant they could sell it for less.

Clothing Production

Mass production also made clothing more plentiful and affordable. For many centuries, cloth and

Supermarkets open across
the United States.
1920s

Refrigerators with freezer
compartments are sold.
1940s

A woman brings spun threads to a weaver who weaves it into cloth. The cloth, which could only be as wide as the loom, was then pieced together into clothing.

clothing were made by hand. Wool, linen, and cotton were common fabrics. Raw wool comes from sheep, and cotton and linen flax are grown as crops. After the sheep were sheared or the crops were picked, families cleaned and combed the raw materials. Spinners (usually women) then twisted raw wool and linen flax into thread on spinning wheels. Weavers (usually men) wove the threads into cloth on small looms.

Looms could not be any wider than a man's arm span, because the weaver had to pass the shuttle (thread carrier) back and forth between his hands. Weaving was hard work, and the pieces of woven cloth were not very large. Then, in 1733, John Kay, from Lancashire, England, invented a machine called the flying shuttle. When the weaver pulled the cord on a mechanical piece called a driver, the driver shot the flying shuttle across the loom.

With a flying shuttle machine, a weaver could do twice as much work in the same amount of time. But spinners still did their work by hand, and they couldn't supply enough thread to keep the weavers busy. In 1764 James Hargreaves, a weaver from England, found a solution. He invented a spinning machine that spun thread on several spindles (wooden pins) at once. It did the work of eight spinners. Hargreaves named the machine the spinning jenny after his daughter.

The flying spindle and the spinning jenny machines were still small enough to be used in weavers' and spinners' homes. But larger spinning machines and weaving looms were also built in textile mills (or cloth factories) across England. In 1790 the first U.S. cotton mill opened in Pawtucket, Rhode Island. Soon mills sprang up in Massachusetts, Connecticut, New Hampshire, and New York. Regions in England and New England rapidly changed from farming communities to mills and mill towns.

In mills, cloth was produced faster and in larger quantities than ever before. As a result, it was cheaper and people could afford to buy more. But most people still had to sew the cloth into garments

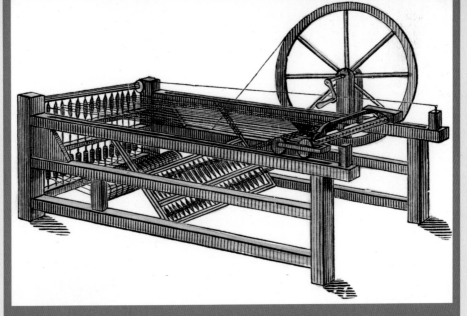

The spinning jenny's wheel spun thread onto many spindles (along the back of the machine) at the same time.

themselves. Every seam, cuff, and hem had to be stitched by hand. Of course, people could buy clothes from tailors if they had enough money. These expensive clothes were made individually to fit the person who bought them. Most ready-made clothing—clothes sewn to fit anyone rather than tailored—was intended for poor people and slaves.

The Sewing Machine

Then, in 1846, an American named Elias Howe patented the first sewing machine. This invention allowed tailors and regular people who could afford a machine to make clothes ten to fifteen times

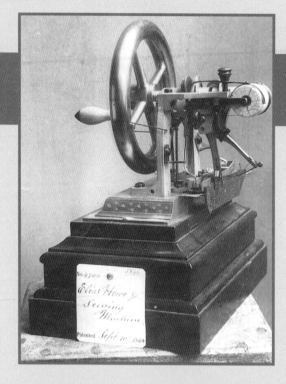

Elias Howe's sewing machine was patented in September 1846.

faster than they could by hand. In the 1860s, techniques for cutting cloth were improved, allowing workers to cut more cloth for clothes faster. These advancements made it possible for companies to make ready-made clothes of high quality. They were no longer just for the poor.

Following the Civil War, ready-made clothing was manufactured for women as well as men. By 1910 sewing and cutting machines were powered by electricity, making clothes production even faster. Companies could make large amounts of clothing in factories. As with food, making a large quantity allowed companies to sell the

Women model ready-made dresses at the John Wanamaker department store in Philadelphia, Pennsylvania, in 1917.

product for less. By World War I, most Americans had made the switch from hand-made clothing to factory-made clothing.

The world had changed forever. People no longer had to labor the entire day just to meet their most basic needs. Our lives also gained fantastic variety. You could have a different meal every day for a month and there'd still be countless foods you hadn't tried. You can pick garments in different styles, fabrics, and colors. There is something for just about everyone's size and taste. Mass production makes it possible.

The White Company begins selling a computerized sewing machine.
1970s

CHAPTER 5

Microwave Ovens

You're hungry and in a hurry. You need a meal in minutes. Do you have to settle for cold leftovers? Should you just grab some junk food? No, you're in luck—the microwave oven has been invented!

About 90 percent of U.S. homes have microwave ovens. Most young people have grown up with these appliances. You might not be able to picture your life without one. Yet microwave ovens are a fairly new invention. Before they existed, people often spent hours cooking the same meals they now make in minutes.

Until recently, housewives were usually responsible for food preparation. Often they began planning meals days in advance. To have everything done by dinnertime each evening, they often started cooking right after lunch. Preparing food for a whole family took up a good part of their day.

A Melted Candy Bar

Then, in the mid-1940s, home cooking changed dramatically. A self-taught engineer named Percy Lebaron Spencer was working at his job at the Raytheon Corporation in Waltham, Massachusetts. Raytheon was an engineering company that often made equipment for the Defense Department. The United States was fighting World War II (1939–1945), and development of defense equipment was a high priority.

Spencer was testing a new power tube called a magnetron. The magnetron produced short, high-frequency radio waves called microwaves. The U.S. military had begun using microwaves in radar systems for detecting enemy airplanes.

After testing the new magnetron, Spencer reached into his pocket for a candy bar he had brought to work. He was shocked to find that it had melted. The room wasn't hot, and he wasn't standing next to any obvious heat sources. Even though the magnetron didn't feel hot to the touch, Spencer suspected that the microwaves had melted the chocolate.

Microwaves melt Percy L.
Spencer's candy bar.

1945

Spencer did some more experiments with microwaves and food. He tested the effect of microwaves on popcorn kernels and on an egg. Sure enough, the popcorn popped, and the egg burst open. Spencer had accidentally stumbled upon a new method for rapid cooking! In 1946 a group of Raytheon engineers patented the microwave cooking process.

Microwave ovens cook food differently from conventional ovens. When the oven is turned on, the magnetron produces microwaves. These waves bounce back and forth inside the oven's metal interior until they are absorbed into the food. The microwaves cause water molecules in the food to vibrate rapidly. The vibrations cause friction, which produces the heat that cooks the food.

NOT JUST FOR COOKING

While microwaves are most commonly used in microwave ovens, they have also been used:

- To detect speeding cars
- To send telephone and television communications
- To dry potatoes to make potato chips
- To roast coffee beans and peanuts
- To dry paper, plywood, leather, tobacco, flowers, and other items

The Radarange

The first microwave oven was ready for use by 1947. It was about the size of a refrigerator. By the following year, the Raytheon Corporation had manufactured a number of these models. They were called

Raytheon engineers
patent their microwave
cooking process.

1946

A woman in 1947 demonstrates the super-fast cooking ability of the Radarange!

Radaranges and were marketed and sold to be used in restaurants.

These first microwave ovens were not suitable for home use. Besides their large size, they cost several thousand dollars. But the ovens helped restaurants to serve both fresh and frozen foods more quickly. Quick meals were especially helpful to people on their lunch breaks who had to get back to work quickly.

By the early 1950s, new microwave units were introduced for home use. These ovens were small enough to be placed on a countertop.

Microwave ovens are
sold for home use.
1950s

They were also more reasonably priced. Families could buy these units for less than $500.

The Microwave Boom

During the 1970s, the demand for microwave ovens shot up. Everyone seemed to want the ease and convenience these new appliances offered. By 1975 more Americans bought microwave ovens than standard gas ranges. Modern microwave ovens come with different settings for cooking different foods of different sizes, shapes, and types. They are used in countries all over the world.

Microwave ovens have had a tremendous impact on society. Well-balanced meals for whole families can be prepared in minutes. Thousands of companies put them in their staff lunchrooms. People have hot food on their work breaks and lunch hours.

UNIVERSAL APPEAL

More than 200 million microwave ovens are used around the globe.

Think of the time microwave ovens save you. A baked potato takes about an hour in a regular oven, but it cooks in a microwave oven in about seven minutes. A frozen chicken breast needs several hours to defrost naturally. But a microwave oven can do the job in about five minutes. Have you ever popped popcorn in a pan on a stovetop? It takes about twenty minutes and involves hot, messy oil, not to mention washing the pan. It's a lot easier to toss a bag of

Microwave ovens outsell standard gas ranges.

1975

A woman in her kitchen prepares food in her microwave oven. The microwave has become a standard appliance in most modern-day homes.

microwave popcorn into a microwave oven for a few minutes.

In the early years of microwave ovens, many working mothers claimed that their microwave was a lifesaver. Most modern families don't even think about it. They can't even imagine life without one.

Ninety percent of
U.S. homes have
microwave ovens.
2000

Epilogue

The microwave ovens of tomorrow will make cooking even easier. The Microsoft Corporation offers a glimpse of the future in a model home it has built in Redmond, Washington. No one lives in this house, but it is equipped with some nifty devices and appliances. It shows what homes may be like in the future.

Among the futuristic appliances is a microwave oven that reads the bar codes on packages of food. The microwave then automatically sets up the cooking time and other settings. The cook doesn't even have to remain in the kitchen to know when the food is cooked. A personal computer or television in another room can be set to notify the person that their meal is ready.

Other products that will simplify everyday life are coming soon too. One is a refrigerator that can do an inventory of what's inside it. Using that information, the fridge can make a grocery shopping list for you. Another is a furnace that can sense when you're not home and will turn itself down to save energy.

flushing toilet was made about 4,000 years ago in ancient Crete, an island southeast of mainland Greece. There, a queen's bathroom contained a toilet made of a wooden seat, a pan, and a system of pipes and drains that flushed waste from the pan and into the underground sewers.

Europe Slacks Off

In Europe the Roman Empire was responsible for building sewer and aqueduct (water-carrying) systems. After the fall of the Roman Empire (about A.D. 475), few European countries took care of or expanded these systems. Without plumbing and sewer systems, bathing was rare and householders got rid of waste however they could.

Rome's last emperor, Romulus Augustulus, is dethroned.

A.D. 476

By the Middle Ages, waste disposal had become a serious problem in large European cities. In some cities, waste was dumped in the nearest river. This posed a serious health risk, as people often got their drinking water from these same rivers. In other cities, the streets were open sewers. Waste just lying in the streets easily spread germs and bacteria. Serious diseases such as typhoid fever, cholera, and dysentery became common.

Some people recognized the need for cleaner, more convenient waste disposal. The person most often credited with inventing the modern flushing toilet is Sir John Harington, a godson of Queen Elizabeth I of England. In 1596 Harington designed a toilet, or "water closet," as it was called, for the queen. The queen's toilet was installed at Richmond Palace, and Harington made a similar one for himself as well. Both the queen and her godson used these devices for years. However, Harington wrote a book in which he joked about his invention, and the book gave people a negative opinion of water closets. Indoor toilets didn't catch on with common people right away.

Sir John Harington

John Harington designs a
"water closet."

1596

Home safety is another area being considered in designing new houses. Home computers would allow people to watch over their homes even while away on vacation. Computers would also link us to the homes of family or friends. If, for example, an elderly relative in another city doesn't pick up the day's mail or turn the lights on at night, it could be a signal to call to make sure nothing is wrong. Other monitors in the home could test for air-quality problems such as carbon monoxide (a deadly, invisible gas), smoke from a fire, or mold.

These products are only a few years away. Before long, they will seem as common as washing machines and store-bought broccoli. Even more advancements will be made in the future. Inventors are always coming up with new ideas. These new appliances may give us more free time in the future. The best may be yet to come.

CA. 3000 B.C. The first-known indoor bathrooms are used on the Orkney Islands, near Scotland.

A.D. 1250s Fireplaces replace indoor fire pits in some European castles.

1300s In Rome, chamber pots are regularly emptied by tossing the contents out into the street.

1500s Chimneys with fireplaces help funnel smoke out of houses.

1596 Sir John Harington designs a toilet for Queen Elizabeth I of England.

1797 The washboard is invented.

1803 The first icebox patent is issued.

1810 Nicolas Appert of France receives the first patents for canning.

1813 The British navy begins to can food in tin cans rather than glass jars. Tin cans are cheaper and don't break easily.

1827 An ice cutter is invented, allowing Americans to efficiently cut ice from ponds and lakes, greatly reducing the cost of ice.

1846 Elias Howe patents his first sewing machine in America.

1849 The California gold rush begins, and numerous Chinese laundries open to wash the miners' clothes.

1850s English doctor John Snow proves that polluted drinking water causes cholera.

1857 Julius Adams is hired to design a sewer system for Brooklyn, New York.

1860 Five million cans of food are produced in the United States.

CA. 1864 Ready-made clothes for women are manufactured in the United States.

1870 Thirty million cans of food are produced in the United States.

1870s Refrigerated railroad lines bring fresh food to consumers all over the United States. H. J. Heinz begins to bottle and sell horseradish and pickles.

1874 William Blackstone of Indiana creates a washing machine as a birthday gift for his wife.

1875 The number of washing machine patents worldwide reaches 2,000.

1880 One hundred and twenty million cans of food are produced in the United States.

1880s The Franco-American Food Company begins producing canned meals in New Jersey.

1897 Condensed canned soup is invented in New Jersey.

1908 The first electric washing machine is introduced in Chicago, Illinois.

1911 The General Electric Company unveils two of the first home refrigerators in Fort Wayne, Indiana.

CA. 1914 By World War I, ready-made garments are common in the United States.

1915 The first electric clothes dryers hit the scene in the United States.

1920s Supermarkets dot America's landscape.

1922 Maytag becomes the first company in the United States to manufacture agitator washing machines.

1939–1945 World War II. Microwaves are used in radar equipment.

1940s Refrigerators with separate freezer and ice tray compartments are available in the United States.

1945 U.S. engineer Percy L. Spencer discovers that microwaves can be used for cooking.

1947 The Radarange, the first microwave oven in the world, is produced by the Raytheon Corporation in Massachusetts.

1950s Countertop microwave ovens are developed for home use worldwide.

1960s Automatic washing machines become popular worldwide.

2000s Ninety percent of U.S. homes have microwave ovens.

GLOSSARY

aqueduct: a system for carrying a large amount of flowing water

automatic: a machine that operates mostly on its own rather than requiring people to power or guide it

canning: a process for preserving food by heating it and then sealing it in sterilized (germ free) containers

central heating: a method of heating buildings in which heat starts in one place (such as in a furnace) and flows into another system (such as air pipes) that brings it to individual rooms

chamber pot: a small container kept in a room, usually a bedroom, for human waste disposal

chimney: a structure with a channel or tube down the center for drawing or carrying smoke out of indoor spaces

fireplace: a source of indoor heat in which a fire is burned on a hearth and smoke is drawn outside by a chimney

mass production: a system for making items, such as food or automobiles, in large quantities by using automation, continuous process, and other methods

microwave: a form of energy used in radar, communications, cooking, and some factory production

outhouse: a small shelter built outside a main house, used as a toilet

patent: a claim to legal rights for an invention

ready-made: clothing or other items prepared in advance and ready to use by purchasers

sewer: a system for carrying off waste, dirty water, and rainfall

textile mill: a factory for the mass production of cloth

SELECTED BIBLIOGRAPHY

Cowan, Ruth Schwartz. *More Work for Mother: The Ironies of Household Technology from the Open Hearth to the Microwave.* New York: Basic Books, 1983.

DeBonneville, Francoise. *The Book of the Bath.* New York: Rizzoli, 1998.

Feenberg, Andrew. *Questioning Technology.* Florence, KY: Routledge, 1999.

Horan, Julie L. *The Porcelain God: A Social History of the Toilet.* Secaucus, NJ: Citadel Trade, 1997.

Innes, Miranda. *The Fireplace Book: Designs for the Heart of the Home.* New York: Viking Studio, 2000.

Johnson, Duane. *How a House Works.* Pleasantville, NY: Reader's Digest Association, 1994.

Reyburn, Wallace. *Flushed with Pride: The Story of Thomas Crapper.* Englewood Cliffs, NJ: Prentice Hall, 1971.

Seymour, John. *Forgotten Household Crafts: A Portrait of the Way We Once Lived.* New York: Knopf, 1987.

Strasser, Susan. *Never Done: A History of American Housework.* Philadelphia: Owl Books, 2000.

Stratton, Joanna L. *Pioneer Women: Voices from the Kansas Frontier.* New York: Simon & Schuster, 1981.

FURTHER READING AND WEBSITES

"American Inventors and Inventions." *Smithsonian Institution.*
http://www.150.si.edu/150trav/remember/amerinv.htm
This gallery of inventors and inventions has been compiled by the
Smithsonian Institution. Be sure to click on the sewing machine link to read
about Isaac Merritt Singer and his work on the earliest sewing machines.

Colman, Penny. *Toilets, Bathtubs, Sinks and Sewers.* New York: Atheneum,
1994.
Colman traces the history of the bathroom from ancient civilizations to the
present day.

"Cooking Safely in the Microwave Oven." *Food Safety and Inspection Service.*
http://www.fsis.usda.gov/OA/pubs/fact_microwave.htm
Check out this website to learn important safety tips for using your microwave oven.

Erlbach, Arlene. *The Kids' Invention Book.* Minneapolis: Lerner Publications
Company, 1997.
This book shows young readers how to go about creating their own inventions. Information is given on patents as well as on contests for young
inventors.

Jones, Charlotte Foltz. *Accidents May Happen: Fifty Inventions Discovered by
Mistake.* New York: Delacorte, 1996.
A fun text showing how 50 great inventions came about because of accidents. Readers will learn some surprising facts about telephones, matches,
and ice cream sodas.

Knapp, Zondra. *Super Invention Fair Projects.* Los Angeles: Lowell House,
2000.
This book provides everything young readers need to know to design their
own invention fair projects. Information is also given on famous inventors
and their inventions.

Patent, Dorothy Hindshaw. *Homesteading: Settling America's Heartland*. New York: Walker, 1995.
Patent provides a revealing look at what pioneers on the American frontier faced. Readers see what life was like without many of the inventions we now take for granted.

PBS. "Frontier Life" *Frontier House*.
http://www.pbs.org/wnet/frontierhouse/frontierlife/index.html
To accompany its *Frontier House* television series, PBS collected a series of articles on life in the American West. How people built and furnished their houses, how they found food, and how they entertained themselves are all discussed.

PBS. "Levi Strauss." *New Perspectives on the West*.
http://www.pbs.org/weta/thewest/people/s_z/strauss.htm
Levi Strauss was a pioneer in bringing ready-made clothing to the public. Visit this website to learn about the man who helped make jeans a part of American culture.

Ritchie, David. *Frontier Life*. Broomall, PA: Chelsea House, 1996.
The hardships that frontier settlers dealt with are detailed in this interesting history. Readers will enjoy actual photographs from that period.

Rubin, Susan Goldman. *Toilets, Toasters, and Telephones: The How and Why of Everyday Objects*. San Diego: Browndeer Press, 1998.
This informative book provides a history of many of the everyday objects that make our lives easier. It offers a fascinating look into the world of industrial design.

Whitman, Sylvia. *What's Cooking? The History of American Food*. Minneapolis: Lerner Publications Company, 2001.
Whitman tells a detailed history of the foods that Americans have planted, harvested, packaged, and prepared, from the early 1600s to the start of the new millennium.

INDEX

Chapter Opener Photo Captions

Cover Top: Two women wash cloths by hand in the 1800s.

Bottom: A boy helps his mother load a modern washing machine.

pp. 4–5 A woman scrubs clothes on a washboard in the early 1900s.

p. 6 Most early castles, such as this one is Spain, were built for military defense, not comfort. They were cold and dark.

p. 14 A detail from an Italian painting shows a street scene in the Middle Ages.

p. 22 A woman in the early 1800s holds her baby while stirring a pot on the fire. A washtub and washboard are behind her.

p. 29 A woman tends her flock of chickens in the 1870s.

p. 40 Preparing food was once an all-day affair. In a colonial kitchen, a woman tends to food on the fire while another rolls out dough. A third woman in the background churns butter.

pp. 46–47 A woman demonstrates a touch-screen surface kitchen table for accessing the Internet.

About the Author

Award-winning author Elaine Landau worked as a newspaper reporter, a book editor, and a librarian before becoming a full-time writer. She has written more than two hundred nonfiction books for young readers, including *Friendly Foes, Ferdinand Magellan,* and *The History of Energy.* Landau has a bachelor's degree in English and journalism and a master's degree in library and information science. She lives in Miami, Florida, with her husband, Norman, and her son, Michael.

Photo Acknowledgments

The images in this book are used with the permission of: Minnesota Historical Society, pp. 4–5, 29; © Jose Fuste Raga/CORBIS, p. 6; © Historical Picture Archive/CORBIS, p. 8; © Brown Brothers, pp. 10, 26, 31, 38; © Bettmann/CORBIS, pp. 11, 18, 22, 33, 40, 43; © Annie Griffiths Belt/CORBIS, p. 13; © David Lees/CORBIS, p. 14; © Pocumtuck Valley Memorial Association, Memorial Hall Museum, Deerfield MA, p. 15; © Adam Woolfitt/CORBIS, p. 16; © Roger Wood/CORBIS, p. 17; Library of Congress, p. 25 (LC-USZc4-5759); © Brownie Harris/CORBIS, p. 28; © North Wind/North Wind Picture Archives, p. 37; © Schenectady Museum; Hall of Electrical History Foundation/CORBIS, p. 39; © Scott Roper/CORBIS, p. 45; © Panasonic Center/Handout/Reuters/Corbis, pp. 46–47. Front cover: top, © Bettmann/CORBIS; bottom, © Royalty–Free/CORBIS. Montgomery Ward & Co., back cover, p. 1, all borders.